The Great Barrier Reef: The

By Charles River Editors

Picture of a giant clam on the reef

About Charles River Editors

Charles River Editors provides superior editing and original writing services across the digital publishing industry, with the expertise to create digital content for publishers across a vast range of subject matter. In addition to providing original digital content for third party publishers, we also republish civilization's greatest literary works, bringing them to new generations of readers via ebooks.

Sign up here to receive updates about free books as we publish them, and visit Our Kindle Author Page to browse today's free promotions and our most recently published Kindle titles.

Introduction

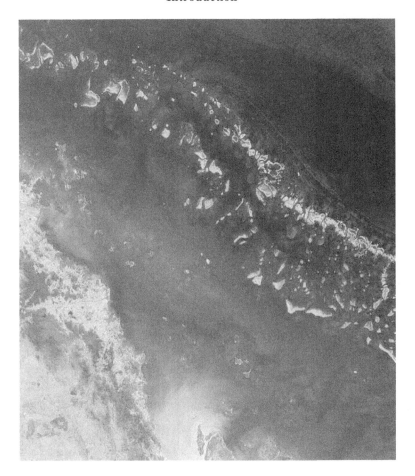

A satellite image of part of the reef

The Greater Barrier Reef

"Coral is a very beautiful and unusual animal. Each coral head consists of thousand of individual polyps. These polyps are continually budding and branching into genetically identical neighbors." – Antony Garrett List

People have always loved to build things, whether it's a feat of engineering in an underground subway or the construction of the world's tallest skyscraper. Thus, it's somewhat ironic that the largest structure ever built was not made by humans but by incredibly tiny organisms known as coral polyps. Over the course of tens of thousands of years, these small organisms have put together a collection of nearly 3,000 reefs that form a collective stretching across 130,000 square miles. It is often mistakenly claimed that the Great Wall of China can be seen in space, but it's absolutely true that the enormous Great Barrier Reef is visible.

The sheer size of the Great Barrier Reef is mind-boggling, but its importance extends far past its physical extent. Put simply, the Great Barrier Reef is one of the most beautiful spots on the planet, offering kaleidoscopic colors thanks to the coral and the species that call it home. This is understandable because a staggering number of species inhabit the Great Barrier Reef, ranging from starfish and turtles to alligators and birds. Scientists have counted about 1,500 different fish species using the reef, and it's estimated that even 1.5 million birds use the site. In designating it a World Heritage Site, UNESCO wrote of the Great Barrier Reef, "The Great Barrier Reef is a site of remarkable variety and beauty on the north-east coast of Australia. It contains the world's largest collection of coral reefs, with 400 types of coral, 1,500 species of fish and 4,000 types of mollusc. It also holds great scientific interest as the habitat of species such as the dugong ('sea cow') and the large green turtle, which are threatened with extinction."

Unfortunately, an ecosystem as complex as the Great Barrier Reef is also vulnerable to a host of threats, whether it's fishing, oil spills, or climate change. J.E.N. Veron, former chief scientist of the Australian Institute of Marine Science, described watching how coral was affected during what's known as a mass bleaching event: "And then I saw a whammy, a mass bleaching event … where everything turns white and dies. Sometimes it's only the fast-growing branching corals, but some of the others are horrible to see; corals that are four, five, six hundred years old—they die, too… It's real, day in, day out, and I work on this, day in, day out. It's like seeing a house on fire in slow motion…There's a fire to end all fires, and you're watching it in slow motion, and you have been for years." In fact, scientists fear that the Great Barrier Reef has lost most of its coral cover in the last 30 years, which poses a danger to the species that inhabit it, some of which are already endangered.

The Great Barrier Reef: The History of the World's Largest Coral Reef looks at the history of the reef and describes it in vivid detail. Along with pictures of important people, places, and events, you will learn about the Great Barrier Reef like never before, in no time at all.

The Great Barrier Reef: The History of the World's Largest Coral Reef
About Charles River Editors
Introduction
   Chapter 1: Freakish Phenomenon
   Chapter 2: A Surge of Romantic Delight
   Chapter 3: The Geologist
   Chapter 4: The Remote Marine Frontier of Torres Strait
   Chapter 5: That Very Partnership
   Bibliography

Chapter 1: Freakish Phenomenon

An aerial view of part of the Great Barrier Reef

"James Cook did not know, on Sunday May 20, 1770, two weeks after leaving Botany Bay on the east coast of New Holland, the western portion of the continent, named by the Dutch captain Abel Tasman in 1644, that the HMS Endeavour was sailing into the southwest entrance of a vast lagoon where reef-growing corals began their work. It was a channel that later navigators would call the Great Barrier Reef inner passage. Cook didn't realize that then, and he never would. The point, obvious enough in his journals, needs stressing because so many historians inadvertently treat this phase of Cook's first voyage of exploration to the Southern Hemisphere as if the Great Barrier Reef we know today already existed somewhere in the back of his mind. As if he unconsciously knew he was about to enter into combat with a constellation of deep-water 'barrier reefs' that ran more or less parallel with the Australian coast for some 1,400 miles, creating between them and the mainland a shallow lagoon of uneven depths interspersed with three hundred reef-fringed coral cays and striated with sand, rock, and coral shoals. In reality he sailed unknowingly within the reef lagoon for around 500 miles before he became aware of something resembling a coral 'labyrinth.' Like explorers before him, he'd had no intimation at all of the possible existence of this freakish phenomenon." – Iain McCalman, *The Reef: A Passionate History: The Great Barrier Reef from Captain Cook to Climate Change*

Anthropologist say that the first human beings to live on the islands that make up the reef settled there more than 60,000 years ago, but little is known about these ancient people except what can be learned from the items they left behind. Frankly, then as now, the animals in and around the reef are more interesting than the humans, and when a previously undiscovered species shows up, it makes the news. For example, in 2010, John Marshall, a researcher who has spent much of his career exploring the reef, described "prehistoric six-gilled sharks" he was able to observe in the water deep under the reef and out of the reach of sunlight: "Some of the creatures that we've seen we were sort of expecting, some of them we weren't expecting, and some of them we haven't identified yet. ... There was a shark that I really wasn't expecting, which was a false cat shark, which has a really odd dorsal fin. ... One of the things that we're trying to do by looking at the life in the deep sea is discover what's there in the first place, before we wipe it out. We simply do not know what life is down there, and our cameras can now record the behavior and life in Australia's largest biosphere, the deep sea."

No matter how long these and other similar marine life have inhabited the reef, the recorded history of the area began in 1770 when the famous explorer Captain James Cook explored the waters off the coast of modern day Queensland. On May 20, he recorded in his journal, "Having but little wind all Night, we kept on to the Northward, having from 17 to 34 fathoms, from 4 Miles to 4 Leagues from the Land, the Northernmost part of which bore from us at daylight West-South-West, and seemed to End in a point, from which we discovered a reef stretching out to the Northward as far as we could see, being, at this time, in 18 fathoms; for we had, before it was light, hauled our Wind to the Westward, and this course we continued until we had plainly discovered breakers a long way upon our Lee Bow, which seemed to Stretch quite home to the land. We then edged away North-West and North-North-West, along the East side of the Shoal, from 2 to 1 Miles off, having regular, even Soundings, from 13 to 7 fathoms; fine sandy bottom. At Noon we were, by Observation, in the Latitude of 24 degrees 26 minutes South, which was 13 Miles to the Northward of that given by the Log. The extreme point of the Shoal we judged to bear about North-West of us; and the point of land above-mentioned bore South 3/4 West, distant 20 Miles. This point I have named Sandy Cape on account of 2 very large white Patches of Sand upon it. It is of a height Sufficient to be seen 12 Leagues in Clear weather (Latitude 24 degrees 46 minutes, Longitude 206 degrees 51 minutes West); from it the Land trends away West-South-West and South-West as far as we could see."

Captain Cook

Cook continued to explore the reef and its islands, and a few days later, he wrote, "At 5 A.M. we made sail; at daylight the Northernmost point of the Main bore North 70 degrees West, and soon after we saw more land making like Islands, bearing North-West by North; at 9 we were abreast of the point, distant from it 1 mile; Depth of Water 14 fathoms. I found this point to lay directly under the Tropic of Capricorn, and for that reason call it by that Name. ... It is of a Moderate height, and looks white and barren, and may be known by some Islands which lie to the North-West of it, and some small Rocks one League South-East from it; on the West side of the Cape there appeared to be a Lagoon. On the 2 Spits which form the Entrance were a great Number of Pelicans; at least, so I call them. The most northernmost land we could see bore from Cape Capricorn North 24 degrees West, and appeared to be an Island."

On June 11, Cook and his men became the first people in recorded history to damage the delicate reefs, though to be fair, the reef did plenty of damage to his ship as well. He described the incident: "Having the advantage of a fine breeze of wind, and a clear Moon light Night in standing off from 6 until near 9 o Clock, we deepened our Water from 14 to 21 fathoms, when all at once we fell into 12, 10 and 8 fathoms. At this time I had everybody at their Stations to put about and come to an Anchor; but in this I was not so fortunate, for meeting again with Deep Water, I thought there could be no danger in standing on. Before 10 o'clock we had 20 and 21 fathoms, and continued in that depth until a few minutes before 11, when we had 17, and before the Man at the Lead could heave another cast, the Ship Struck and stuck fast. Immediately upon this we took in all our Sails, hoisted out the Boats and Sounded round the Ship, and found that we had got upon the South-East Edge of a reef of Coral Rocks, having in some places round the Ship 3 and 4 fathoms Water, and in other places not quite as many feet, and about a Ship's length from us on the starboard side (the Ship laying with her Head to the North-East) were 8, 10, and 12 fathoms. As soon as the Long boat was out we struck Yards and Topmast, and carried out the Stream Anchor on our Starboard bow, got the Coasting Anchor and Cable into the Boat, and were going to carry it out in the same way; but upon my sounding the 2nd time round the Ship I found the most water a Stern, and therefore had this Anchor carried out upon the Starboard Quarter, and hove upon it a very great Strain; which was to no purpose, the Ship being quite fast, upon which we went to work to lighten her as fast as possible, which seemed to be the only means we had left to get her off."

A replica of Cook's ship, the HMS *Endeavour*

A contemporary illustration depicting the ship

Looking back on that fateful day, the ship's physician, Dr. John Hawkesworth, recalled, "Hitherto we had safely navigated this dangerous coast, where the sea in all parts conceals shoals that suddenly project from the shore, and rocks that rise abruptly like a pyramid from the bottom, for an extent of two and twenty degrees of latitude, more than one thousand three hundred miles; and therefore hitherto none of the names which distinguish the several parts of the country that we saw, are memorials of distress; but here we became acquainted with misfortune, we therefore called the point which we had just seen farthest to the northward, Cape Tribulation."

For hours, the men worked to removed everything they could from the stuck ship, hoping that the lightened vessel would be able to float off her position, but this was not to be. Cook continued, "As we went ashore about the Top of High Water we not only started water, but threw overboard our Guns, Iron and Stone Ballast, Casks, Hoop Staves, Oil Jars, decayed Stores, etc.; many of these last Articles lay in the way at coming at Heavier. All this time the Ship made little or no Water. At 11 a.m., being high Water as we thought, we tried to heave her off without Success, she not being afloat by a foot or more, notwithstanding by this time we had thrown overboard 40 or 50 Tons weight. As this was not found sufficient we continued to Lighten her by every method we could think off; as the Tide fell the ship began to make Water as much as two pumps could free: at Noon she lay with 3 or 4 Streakes heel to Starboard; Latitude observed 15 degrees 45 minutes South."

One of the ship's cannons was later recovered.

The men and their Captain worked through the night, and the next day Cook confided to his journal, "Fortunately we had little wind, fine weather, and a smooth Sea, all this 24 Hours, which in the P.M. gave us an Opportunity to carry out the 2 Bower Anchors, one on the Starboard

Quarter, and the other right a Stern, got Blocks and Tackles upon the Cables, brought the falls in abaft and hove taught. By this time it was 5 o'Clock p.m.; the tide we observed now begun to rise, and the leak increased upon us, which obliged us to set the 3rd Pump to work, as we should have done the 4th also, but could not make it work. At 9 the Ship righted, and the Leak gained upon the Pumps considerably. This was an alarming and, I may say, terrible circumstance, and threatened immediate destruction to us. However, I resolved to risk all, and heave her off in case it was practical, and accordingly turned as many hands to the Capstan and Windlass as could be spared from the Pumps; and about 20 Minutes past 10 o'Clock the Ship floated, and we hove her into Deep Water, having at this time 3 feet 9 Inches Water in the hold. This done I sent the Long boat to take up the Stream Anchor, got the Anchor, but lost the Cable among the Rocks; after this turned all hands to the Pumps, the Leak increasing upon us."

After repairing his ship, Cook continued exploring the area, but with much greater care. On June 30, he jotted down, "At this time it was low water, and I saw what gave me no small uneasiness, which were a Number of Sand Banks and Shoals laying all along the Coast; the innermost lay about 3 or 4 Miles from the Shore, and the outermost extended off to Sea as far as I could see without my glass, some just appeared above water. The only hopes I have of getting clear of them is to the Northward, where there seems to be a Passage, for as the wind blows constantly from the South-East we shall find it difficult, if not impractical, to return to the Southward."

A century and a half later, author William Seville Kent explained what had happened: "Captain Cook was not aware of it; but there were, in the near vicinity of Lizard Island, two passages through the Barrier far more practicable than the one he penetrated. One of these, known as the One and a Half Mile Opening, is less than ten miles north of Cook's Passage, and the other, the Lark Pass, just forty miles south of the same point. None of the three passages, nor, indeed, any other that penetrates the Barrier farther north, is to the wide, open character that characterizes the channels and openings to the south, previously enumerated; and it is significant in association with this phenomenon, that no large rivers, draining a considerable extent of back country, fall upon the northern side of the eastern coast. Such larger rivers as do exist flow westward to the Gulf of Carpentaria. At the same time, some correlation might possibly be established between the Endeavour River estuary and the Lark Passage, and between the estuary of Kennedy River and the First Three Mile opening, a little to the north of Cape Melville. In both instances the Barrier gaps lay some little distance to the north of the rivers' mouths, and the connection between the two is consequently not so obvious as in the examples previously recorded."

Lizard Island

Photo of parts of the reef from a helicopter

Chapter 2: A Surge of Romantic Delight

"In September and early October 1802 Flinders knew them only as the start of Cook's notorious labyrinth. Having charts made by Cook, Campbell, and Swain didn't actually speed progress. Flinders was still forced to adopt Cook's painstaking sailing regime of anchoring at night, taking continual soundings, and positioning lookouts on the forecastle and topmast. Several attempts to thread the Investigator through narrow channels in the coral saw the ship driven back by tides, whirlpools, and currents that swirled in the funnel-like entrances. A frustrated Flinders found himself echoing his predecessor's complaints: 'for by this time I was weary of them [reefs], not only from the danger to which the vessels were thereby exposed, but from fear of the contrary monsoon setting in upon the North Coast, before we should get into the Gulph of Carpentaria.' By October 9 the vessels were so hemmed in by coral that Flinders had to resort to a further Cook expedient. Anchoring until low tide, he set out in a whaleboat to explore the line of exposed reefs in the hope of finding navigable channels. Clambering for the first time onto one of these reefs, he couldn't refrain, even in his present predicament, from expressing a surge of romantic delight at the sight that met his eyes." – Iain McCalman, *The*

Reef: A Passionate History: The Great Barrier Reef from Captain Cook to Climate Change

A generation after Cook had his famous "run in" with the Great Barrier Reef, Captain Matthew Flinders returned to the area, determined to add to what Cook had already recorded about the mysterious navigational "labyrinth." During his time there in the fall of 1802, he accomplished this and so much more, making some of the first notes about the beauty of the reef itself: "In the afternoon I went upon the reef with a party of gentlemen, and the water being very clear round the edges, a new creation, as it was to us, but imitative of the old, was there presented to our view. We had wheat sheaves, mushrooms, stags' horns, cabbage leaves, and a variety of other forms, glowing under water with vivid tints of every shade, betwixt green, purple, brown, and white; equaling in beauty, and excelling in grandeur, the most favorite parterre of the curious florist. There were different species of coral and fungus growing, as it were, out of the solid rock, and each had its peculiar form and shade of coloring; but whilst contemplating the richness of the scene we could not long forget with what destruction it was pregnant. Different corals in a dead state, concreted into a solid mass of a dull white color, composed the stone of the reef. The negro heads were lumps that stood higher than the rest; and, being generally dry, were blackened by the weather; but even in these the forms of the different corals, and some shells were distinguishable. The edges of the reef, but particularly on the outside where the sea broke, were the highest parts; within these were pools and holes containing live corals, Sponges, sea eggs, and cucumbers; and many enormous cockles (chama gigas) were scattered upon different parts of the reef. At low water this cockle seems most commonly to be half open; but frequently closes with much noise, and the water within the shells then spouts up in a stream three or four feet high; it was from this noise and the spouting of the water that we discovered them, for in other respects they were scarcely to be distinguished from the coral rock."

Flinders

Flinders then went on to make some of the first scientific speculations about the origins and growth of the reefs: "It seems to me, that when the animalcules which form the corals at the bottom of the ocean cease to live, their structures adhere to each other, by virtue either of the glutinous remains within, or of some property in salt water; and the interstices being gradually filled up with sand and broken pieces of coral washed by the sea, which also adhere, a mass of rock is at length formed. Future races of these animalcules erect their habitations upon the rising bank, and die, in their turn, to increase, but principally to elevate, this monument of their wonderful labors. The care taken to work perpendicularly in the early stages would mark a surprising instinct in these diminutive creatures. Their wall of coral, for the most parts in situations where the winds are constant, being arrived at the surface, affords a shelter, to leeward of which their infant colonies may be sent forth; and to this their instinctive foresight, it seems to be owing, that the windward side of a reef exposed to the open sea is, generally, if not always, the highest part, and rises almost perpendicular, sometimes from the depth of two hundred, and perhaps many more, fathoms. To be constantly covered with water, seems necessary to the existence of the animalcules, for they do not work, except in holes upon the reef, beyond low water mark; but the coral and other broken remnants thrown up by the sea, adhere to the rock, and form a solid mass with it, as high as the common tides reach. That elevation surpassed, the future remnants, being rarely covered, lose their adhesive property; and remaining in a loose state, form what is usually called a key, upon the top of the reef."

Considering he was working without any serious scientific background, Flinders' assumptions were surprisingly accurate. He went on to describe how the islands among the coral were formed. "The new bank is not long in being visited by sea birds; salt plants take root upon it, and a soil begins to be formed; a cocoa nut, or the drupe of a pandanus is thrown on shore; land birds visit it and deposit the seeds of shrubs and trees; every high tide, and still more every gale, adds something to the bank; the form of an island is gradually assumed; and last of all comes man to take possession. "Half-way island (in Torres strait) is well advanced in the above progressive state; having been many years, probably some ages, above the reach of the highest spring tides, or the wash of the surf in the heaviest gales. I distinguished, however, in the rock which forms its basis, the sand, coral, and shells formerly thrown up, in a more or less perfect form of cohesion; small pieces of wood, pumice stone, and other extraneous bodies, which chance had mixed with the calcareous substances, substances, when the cohesion began, were enclosed in the rock, and in some cases were still separable from it without much force. The upper part of the island is a mixture of the same substances in a loose state, with a little vegetable soil, and is covered with the casuarina, and a variety of other trees and shrubs, which give food to paroquets, pigeons, and some other birds; to whose ancestors, it is probable, the island was originally indebted for this vegetation."

Ultimately, Flinders accurately concluded "that with the exception of…perhaps several small openings, our Barrier Reefs are connected with the Labyrinth of captain Cook; and that they reach to Torres' Strait and to New Guinea … through fourteen degrees of latitude and 9 degrees of longitude; which is not to be equaled in any other known part of the world." He later added, "An arm of the sea is enclosed between the barrier and the coast, which is at first 25 or 30 leagues wide; but is contracted to 20, abreast of Broad Sound, and to 9 leagues at Cape Grenville; from whence it seems to go on diminishing, till, a little beyond Cape Tribulation, reefs are found close to shore. Numerous islands lie scattered in this enclosed space; but so far as we are acquainted, there are no other coral banks in it than those by which some of the islands are surrounded; so that being sheltered from the deep waves of the ocean, it is particularly well adapted to the purposes of a coasting trade. … The commander who proposes to make the experiment [of exploring the reef], must not…be one who throws his ship's head round in a hurry, so soon as breakers are announced from aloft; if he do not feel his nerves strong enough to thread the needle…amongst the reefs, whilst he directs the steerage from the mast head, I would strongly recommend him not to approach this part of New South Wales."

Others who sailed in the Great Barrier Reef also expressed their concerns, with 19th century explorer John Curtis writing a grossly exaggerated and racist complaint: "The Straits of Torres…seem really as if they were destined to be the terror of navigators. This arises from the extreme difficulty of steering through that perilous passage, the irregular courses of the tides, the sudden manner in which storms and hurricanes arise, and the numerous shoals which are scattered in this vast expanse of water seem to bid defiance to nautical skill, and the steadiest caution. To detail the various wrecks which have happened there, that have come to our knowledge, would fill a large folio , and many a vessel has, doubtless, foundered, and been swallowed up in that insatiate gulf, of the particulars of which the world will ever remain ignorant. It is not unlikely that the sanguinary character of the natives, who massacre the survivors who fall into their hands, is the most plausible reason which can be assigned why the fates of many other hopeless vessels are never made known."

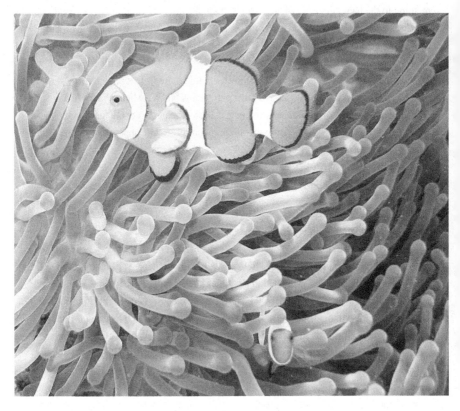

A picture of clownfish among the reef

Picture of a blue starfish resting on coral

Picture of a crown-of-thorns starfish

Chapter 3: The Geologist

"Joseph Jukes, whom Griffin nicknamed 'the geologist,' was officially charged with investigating the geological character of the Great Barrier Reef and the structure, origins, and behavior of reef-growing corals— the first scientist ever to be specifically assigned such a task. Naturally the Admiralty's concern was more practical than scholarly. By the 1840s it was widely recognized that corals were not inert rocks but living organisms, although little was known about the cause, extent, and speed of their development. It was thought that dangerous new reefs might suddenly appear in places where previous surveys had shown nothing. The Admiralty hydrographer Francis Beaufort urged the Fly's Captain Blackwood to remember that he would be dealing with submarine obstacles 'which lurk and even grow.' It was also expected that a geologist would offer expert advice on suitable sites for future harbors and settlements, and when Captain Blackwood gave Jukes responsibility for producing the official journal of the voyage, the geologist stressed that he would approach the task as a down-to-earth scientist, conveying 'plain fact' and 'simplicity and fidelity.' He claimed he would eschew any selecting 'for effect,' or 'heightened recollections,' or 'brilliancy, elegance, or graces of style.'" – Iain McCalman, *The Reef: A Passionate History: The Great Barrier Reef from Captain Cook to Climate Change*

In the early part of 1842, the British government, anxious to know what they had in their new colony, sent the Fly, a navel corvette, to explore the area. Their orders explained not only what they were to do but why they were to do it: "Whereas, a large proportion of the vessels trading to the South Sea, and to Australia, are obliged to return to Europe, or proceed to India, by way of Torres Strait: and whereas many of these vessels, when weak handed, in order to avoid the frequent anchorage necessary in the in-shore passage, by what is called King's Route, stand out to sea till an opportunity offers for making one of the narrow gaps in the Barrier Reefs, through which they steer for the Strait ; and whereas, several vessels have thus been lost, there being no other guide to these openings than the casual observation of latitude which is often incorrect, there being no land to be seen till entangled within the reefs, and no chart on which the dangers are correctly placed. We have therefore thought fit for the above reasons, to have the Great Barrier Reef explored and to have those gaps surveyed, in order that some means may be devised for so marking the most eligible of these openings, that they may be recognized in due time, and passed through in comparative safety."

The admiralty then made it clear that while science would be served by the trip, its primary purpose was one of mapping on behalf of Her Majesty's Fleet. Thus, the Fly's first three orders were:

> "1. The survey of the exterior or eastern edge of that vast chain of reefs which extends almost continuously from Breaksea Spit to the shore of New Guinea.
>
> 2. The thorough examination of all the channels through the Barrier Chain, with detailed plans of those which offer a secure passage.
>
> 3. When you have examined them all, and considered their several advantages and difficulties, and determined which of them will offer the speediest and safest passage for the generality of merchant vessels, you will endeavour to devise some practical means of marking them by beacons of wood, stone, or iron, so placed on their outer islands or cays, that they may serve to guide those vessels to a certain and safe landfall."

On board was naturalist Joseph Beete Jukes, who created the first large complete report on the area. He wrote in part, "I SHALL...give a general sketch of the structure of the Great Barrier reef, as far as it was known at the close of our survey in the year 1846. It may be said to commence with Breaksea Spit, in S. lat. 240° 30', E. long. 158° 20', and extend to Bristow Island, on the coast of New Guinea, in S. lat. 9° 15', and E. long. 148° 20'. This would give, in a straight line, a distance of about 1100 geographical miles...It stretches along the coast at a mean distance of about 30 miles from the land; its outer edge being sometimes not more than 10 or 15, at others, more than 100 miles distant from it. The whole of the sea which lies outside the Barrier, between New Caledonia and Torres Strait, is likewise encumbered with detached reefs of greater or less magnitude. From this large development of coral reefs this sea was called by Flinders the Coral Sea, a name which it well deserves."

Jukes

The government's primary reason for wanting to survey the area was to find better routes through the reef for the increasing number of ships that were by then plying the water along the coast of Australia. Thus, Jukes dealt with that issue first: "In order to traverse the Coral Sea and Torres Strait with any degree of safety, there are two tracks, which are commonly known by the names of the Inner and Outer Routes. In taking the inner passage vessels enter the Barrier reef at its southern extremity, and run up to the northward along shore, between the reefs and the land. …although often narrow and intricate, it is safe, because there is good anchorage the whole of the way, and the reefs themselves are a perfect shelter from the violence of the sea. The outer route has never yet been regularly surveyed, but is known roughly from its having been traversed by whalers and merchant vessels. … Along this route there is a clear track, from 60 to 100 miles in width, that is free from reefs. But outside of it, on either hand, detached reefs are known to be numerous, and there probably exist many which are unknown. In this outer route the sea is of great and almost unfathomable depth as far as is known; there is consequently no anchorage, and whatever the circumstances, a ship must keep under sail till she come up to the edge of the Great Barrier, and pass through one of its openings into the comparatively shallow and sheltered water inside of it."

In spite of the fact that he was hired by the British Navy, Jukes was still a scientist, and he eventually turned his attentions to studying the reef itself. Indeed, he showed particular interest in how such a magnificent form of nature was created, and his records would provide some of the most detailed information about the Great Barrier Reef during the 19th century. "The size and form of an 'individual coral reef' is perfectly indeterminate; it may be circular, oval, or linear; its surface may vary-from a mere point to an area of many square miles. Those, however, which occupy the extreme edge of a mass of reefs, or rise on one side from great depths, having on the other comparatively shallow water, have generally .a linear form, being three, five or ten miles long, and varying in breadth from one or two hundred yards to perhaps a mile. This seems more especially to be the case when their direction runs across that of the prevailing wind. The individual coral reefs which rise from an equal depth all round, whether that depth be great or small, are more commonly of an oval, circular, or irregular shape, but these are usually much larger when exposed to the wind and surf than in more sheltered situations. To get an idea of the nature and structure of an individual coral reef, let the reader fancy to himself a great submarine mound of rock, composed of the fragments and detritus of corals and shells…. There is a term wanted to express the distinction between an individual reef, unbroken by any deep water-channel, and a group of such reefs. For the latter I am almost tempted to use the word 'reefery;' for the former I have, in this passage, used the expression, 'individual coral reef' is quite flat and near the level of low water. At its edges it is commonly a little rounded off, or slopes gradually down to a depth of two, three, or four fathoms, and then pitches suddenly down with a very rapid slope into deep water…. The surface of this reef, when exposed, looks like a great flat of sandstone with a few loose slabs lying about, or here and there an accumulation of dead broken coral branches, or a bank of dazzling white sand. It is, however, checkered with holes and hollows more or less deep, in which small living corals are growing; or has, perhaps, a large portion that is always covered by two or three feet of water at the lowest tides, and here are fields of corals, either clumps of branching madre pores, or round stools and blocks of maeandrina and astrma, both dead and living."

A picture of the coral *Turbinaria mesenterina*

One of the things that Jukes soon noticed was that the location of each part of the Great Barrier Reef significantly influenced how large it was and how it was shaped. "Proceeding from this central flat towards the edge, living corals become more and more abundant. As we get towards the windward side, we of course encounter the surf of the breakers long before we can reach the extreme verge of the reef, and among these breakers we see immense blocks, often two or three yards (and sometimes much more) in diameter, lying loose upon the reef. These are sometimes within reach by a little wading; and though in some instances they are found to consist of several kinds of corals matted together, they are more often found to be large individual masses of species, which are either not found elsewhere, and consequently never seen alive, or which greatly surpass their brethren on other parts of the reef in size and importance. If we approach the lee edge of the reef, either by wading or in a boat, we find it covered with living corals, commonly maeandrina, astraea, and madre pore, in about equal abundance, all glowing with rich colors, bristling with branches, or studded with great knobs and blocks."

A picture of orange sun coral

Toby Hudson's picture of corals of various colors

Again and again in his exploration, Jukes mentioned the landscape that the reef both created and was created by. "When the edge of the reef is very steep, it has sometimes overhanging ledges, and is generally indented by narrow winding channels and deep holes, leading into dark hollows and cavities where nothing can be seen. When the slope is more gentle, the great groups of living corals and intervening spaces of white sand can be still discerned through the clear water to a depth of forty or fifty feet, beyond which the water recovers its usual deep blue. A coral reef, therefore, is a mass of brute matter, living only at its outer surface, and chiefly on its lateral slopes. It is believed that coral animals cannot live at a great depth; that twenty, or at the most thirty fathoms is their extreme limit of growth. This is apparently proved, or nearly so, with respect to all known species of I have seen a block of maeandrina, of irregular shape, twelve or fifteen feet in diameter, the furrows of which, though much worn and nearly obliterated, were wider than my three fingers; also very large blocks and crags of a porites, twenty feet long and ten feet high, but all one connected mass, without any breaks in its growth together in various ways, forming 'atolls,' 'barriers,' and ' fringing reefs,' as described by Mr. Darwin."

Richard Ling's picture of a lizardfish and sponges on the reef

Richard Ling's picture of a banded coral shrimp

As a scientist, Jukes was better equipped that many of the others who had previously explored the area to observe and classify the different parts of the reef: "The Great Barrier Reef of Australia is itself composed of each of these different kinds of groups, which I shall class as 1st—'The linear reefs,' forming the outer edge, or actual Barrier. 2nd—'Detached reefs,' lying outside that Barrier. 3rd— 'Inner reefs,' or those which lie between the Barrier and the shore. The 1st, or linear reefs, are generally long and narrow, running along a line more or less parallel with the shore, and divided from each other by narrow passages. They vary from 200 yards to a mile in width, and from half a mile to ten or fifteen miles in length. They have commonly unfathomable, or at least: unfathomed, water on the outside, and a depth of from ten to twenty fathoms on the inside of them. The 2nd or 'Detached reefs,' are not common near the Barrier, and occur only in one neighborhood. They rise from deep water all round, have more or less of a circular form with lagoons inside them, and are regular 'atolls' on a small scale. The 3rd or 'Inner reefs,' are very numerous. They are scattered over the space between the Barrier and the land, sometimes occupying the greater part of the intermediate space, sometimes leaving a comparatively clear channel on one or both sides of them; that is between themselves and the Barrier, and between themselves and the land. They have no peculiarity of form, are perhaps most commonly steep-sided, but not unfrequently have a gradual slope."

The southernmost island in the Great Barrier Reef is Lady Elliot, and it was here that Jukes observed some of the flora that was prevalent on the island. "The beach was composed of coarse fragments of worn corals and shells, bleached by the weather. At the back of it a ridge of the same materials, four or five feet high and as many yards across, completely encircled the island, which was not a quarter of a mile in diameter. Inside this regular ridge were some scattered heaps of the same stuff, the whole encircling a small sandy plain. The encircling ridge was occupied by a belt of small trees, while on the plain grew only a short scrubby vegetation, a foot or two high. The materials of the encircling ridge were quite low, and thinly covered with vegetable soil among the trees ; but the sand of the central plain, which was dark brown, was sufficiently compact to be taken up in lumps, and a little underneath the surface it formed a kind of soft stone, with embedded fragments of coral. Some vegetable soil also was found, a few inches in thickness in some places, the result of the decomposition of vegetable matter and birds' dung. On the lee, or north-west side of the island, was a coral shoal or bank, sloping gradually off, from low-water mark for about a quarter of a mile, when it was two or three fathoms under water. Immediately beyond this was a depth of fifteen fathoms."

Pictures of Lady Elliot Island and the surrounding reef

Likewise, Jukes documented several species of life living among the reefs near the island. "On the south-east, or weather side of the island, was a coral-reef about two miles in diameter, having the form of a circle of breakers, including a shallow lagoon. Among the breakers, on the external edge of the reef some large black rocks showed themselves above water here and there all around. The lagoon inside was shoal, having two or three fathoms' water occasionally over spaces of white sand, the rest being occupied by flats of dead and living coral, of which the former was left dry at low water. In this lagoon we saw both sharks and turtle swimming about, and there were upwards of thirty fine turtle 'turned' when the boats first landed. One island was well-stocked with birds, of which black noddies and shearwaters were the most abundant; the next in number being terns, gulls, white herons or egrets, oyster-catchers, and curlews. The trees were loaded with the nests of the noddies, each of which was a small platform of seaweed and earth, fixed in the fork of a branch. They had one rather elongated lightish brown egg, rather less than a hen's egg. The shearwaters burrowed in the ground two or three feet; their eggs were larger, rather pointed and speckled, and streaked with black. On the south side of the island, on the beach, were exposed some beds of pretty hard rock, formed of fragments of corals and shells, compacted together in a matrix of still smaller grains of the same material. The beds were thin and slab-like, and rose from the lagoon at an angle of about $8°$ to a height of six or eight feet above high-water mark. Some of the finer slabs reminded me very much in general appearance of the slabs of the Dudley limestone. The color of the rock was dark brown, hard externally, but the inside was white and much softer."

Richard Ling's picture of a reef stonefish

A picture of Crescent-tail Bigeyes swimming along the reef

Jukes continued to make many notes on the biodiversity of the Great Barrier Reef, paying particular attention to the many different species of coral he encountered. He recorded in his journal the following observations on January 11, 1843: "Landed on this (One-Tree) island, which exhibited the same general features as Bunker's first island, with some modifications. The external ridge of loose coral fragments was loftier and steeper, owing, I believe, to this island being rather more on the weather, or at least the south side of the reef. Inside, the island sloped down every way towards the center, forming a shallow basin, in the middle of which was a small hole of salt-water at or near the level of the sea. The inside slope was covered with low succulent plants with pink flowers (Mesembryanthemum) and low trailing bushes. On his green carpet were multitudes of young terns that fluttered before us like flocks of ducklings, with the old birds darting and screaming over our heads. In the single tree (which was, in fact, a small clump of the common Pandanus of these seas with its roots exposed above ground) was a large rude mass of old sticks, the nest of some bird of prey, probably the osprey."

Aerial view of One-Tree Island

Having completed his work on One-Tree Island, Jukes moved on: "To the northward and eastward of the island stretched the shoal lagoon, its bottom of clean white sand, and dark patches of dead and living coral, bounded by the usual rim of snow white breakers. Just round the island, part of the body of the reef was now exposed at low water. This was a flat surface of about a quarter of a mile in width, dotted here and there with pools and holes of water. It consisted of a compact, tough, but rather soft and spongy rock, many loose slabs of which, two or three inches thick, were lying about. It was rather fine-grained, and only here and there exhibited any organic structure or remains. There were no signs of living coral, except a few stunted specimens in some of the deeper holes of the reef, where also were some dead masses still standing in the position of growth. The whole was very different from any preconceived notions of a coral reef, and I erroneously imagined it must be an exception to their general character; it looked simply like a half-drowned mass of dirty brown sandstone, on which a few stunted corals had taken root."

The next day, Jukes explored what would later be called Heron Island, though he had a rather difficult time going up the beach: "In attempting to land at low water, we were compelled to quit the boat soon after getting on the edge of the reef, and wade ashore a distance of a third of a mile. The bottom was very irregular, but pretty equally, divided between white sand and blocks of dead and living coral, principally the former. On many of the rough blocks of coral there was scarcely a few inches of water, and many large masses, particularly along the outer edge of the reef, were high and dry. All the sandy spots, however, were about three or four feet deep, and as neither the sandy spots nor the coral-masses were anywhere continuous for more than a yard or two, we had a succession of wading and scrambling that was rather laborious."

Aerial photos of Heron Island

Once on the island, he was again fascinated by what he observed, and without thinking he joined the ranks of those who committed unnecessary violence against the reef. He wrote in his journal, "Arrived at the island, the first thing that took my attention was a large development of hard brown rock, like that on Bunker's Island. Both the island and the reef were elongated in an east and west direction, the island being half a mile long, and not more than 300 yards broad. It consisted in the interior of piles of loose sand, covered by a dense wood of pretty large trees, with broadish leaves, most of which had a white brittle wood, and grew in a singularly slanting position, the stems frequently curving at an angle of 45°, although three or four feet in circumference. The beach of the island was steep, about twenty feet high at low water, and composed partly of sand and partly of stone. The sand was very coarse, composed wholly of large grains and small angular pieces of comminuted corals and shells, with some larger worn fragments of both intermixed. The stone was of precisely the same materials, but very hard, and dark brown externally, although still white inside. It sometimes required two or three sharp blows with the hammer to break even a corner of it off. Its surface was everywhere rough, honeycombed and uneven; the beds from one to two feet in thickness, with, occasionally, in the fine-grained parts, a tendency to split into slabs or flags. It was perfectly jointed by rather zig-zag points crossing each other at right angles, and splitting the rock into quadrangular blocks of from one to two feet in the side. As far as external appearance and character went, it might have been taken for any old roughly stratified rock. As to position, the strike of the rock was parallel to the direction of the long diameter of the island and reef or east and west; and it dipped on the north and south sides of the island to the north and south respectively; or from the island towards the reef at an angle of 8° or 10°."

Having broken off samples to look at and study, Jukes continued his hike, and he later recalled, "At the east end of the island it was not visible, but at the west it appeared from under the sand in two places, in one being horizontal, and in the other having a slight flexure or anticlinal line, which ranged also east and west. The rock was in many places much worn by the wash of the breakers, which had also a good deal undermined it in some places, and many blocks had fallen down in a line. The joints were parallel to the dip and strike respectively. The rise and fall of tide here was fourteen or fifteen feet, and at high water the upper part of the rock was just about covered; at low water the reef was dry for a small space all-round the island. Now the question is how or under what circumstances did the loose calcareous sand and fragments become hardened into solid stone, acquire a regular bedding and a jointed structure, and the plane of stratification assume an inclination of 8° or 10°."

As is so often the case with a scientific expedition, Jukes' findings raised as many questions as answers: "If it be supposed that a regular deposition and slope of 8° took place every high tide, and a gradual and successive in duration went on, why does not the same thing take place now? Or why did not the loose sand which composes the greater part of the beach in the same position become consolidated? Permanent springs containing carbonate of lime are, of course, improbable in so small a heap of low sand as the islet is composed of. Either, then, the stratification and consolidation is the result of a gradual deposition beneath the level of low water, in which case a movement of elevation must have taken place, which in so small a spot seems a difficult and gratuitous hypothesis; or else the present structure must have been produced in the interior of a mass of loose sand by the in- filtration of sea or rain water, or some other cause of which we are ignorant. I say in the interior, for had it been on the outside, what was to defend it from the wash of the sea that is now breaking down the hard solid rock, and shifting and washing backwards and forwards the loose sand of which the present beach is composed? After the interior of such a mass of sand had been consolidated, the loose exterior may have been washed away and the solid rock exposed. The speculation concerning the structure of this little island may seem a very unimportant circumstance even to the geologist; but it is not so, as this same rock is found along every beach and on every island among the coral-reefs of Australia, and I believe in other parts of the world also."

Although he inevitably damaged parts of the reef in the process, McCalman also acknowledged Jukes' contributions to studies about the Great Barrier Reef. The historian wrote of the naturalist, "The scientist in him was also elated to discover that this region constituted a major faunal boundary line: 'it was evident that in crossing Torres Strait we were passing from the Australian center of life ... into that of the Indian Archipelago.' The differences between the two regions could be observed in the marine species, which met from opposite oceanic directions. Torres Strait shells, echinoderms, and reef burrowers, for example, were 'generally more brilliant in form and color, than those on the Australian coast.' The much damper climate of New Guinea also prevailed in the Torres Strait, contributing to the creation of rich black soil, 'dank woods and jungles,' and a variety of cultivable edible species, including coconut palms, plantains, yams, taro, and sweet potatoes. ... Joseph Beete Jukes, 'the geologist,' was a pioneer of three major traditions of Barrier Reef thought. He was the first scientific analyst of the Reef's geological origins and coral structures, the first professional-style ethnographer of Indigenous Reef cultures, and the first European writer to appreciate the Reef's distinctive romantic beauties."

At the same time, McCalman points out that Jukes gradually (and understandably) grew fonder of the Great Barrier Reef: "Jukes's romanticism...gradually began to inform his views of the Reef's seascapes and landscapes. When younger, he'd felt a 'passion' for poetry, both as reader and writer, being especially fond of Shelley's natural imagery even after he'd outgrown the poet's fuzzy mysticism. Jukes's letters home showed a keen appreciation of romantic aesthetics, and particularly of the fashionable landscape-art theories of the picturesque and sublime. This is not to say that romantic ideas inflected his geological theory, but he admitted privately, if not in his official Admiralty journal, that he also possessed 'a poetic temperament.' He admired Shakespeare above all other writers, and he favored what he called a 'Saxon style' of writing: 'nervous, strong, picturesque, and expressive.' Using this 'picturesque' style, Joseph Beete Jukes became the first explorer-writer to try to persuade his readers that the Great Barrier Reef possessed a distinctive type of beauty and sublimity. He penned moments of rapture that transcended the everyday hardships of being in Reef country— the baking heat, ferocious green ants, swarming three-inch cockroaches, incessant sandflies, and fever-bringing mosquitoes."

Richard Ling's photos of Anemonefish among the coral

Leonard Low's picture of Amphiprion akindynos among the coral

Chapter 4: The Remote Marine Frontier of Torres Strait

"William also discovered a refuge from worldly cares and his past at the remote marine frontier of the Torres Strait, a place where he could exercise his full range of talents and ease the shackles on his stiff personality. After the setback in Tasmania, he was eager to show off his practical usefulness and economic value as a marine scientist and resource manager. A review of Queensland's fishing industries, plus conversations with McIlwraith and Griffith, suggested an urgent need for what he called "a redemption" of the edible oyster and pearl-shell industries, especially the latter. The Torres Strait pearling industry, normally one of Queensland's leading revenue producers at around $ 350,000 a year, had become so exhausted that much of the harvested shell was now too small for button manufacturers to use. Ever since his time as an aquarium biologist, William had championed artificial cultivation as a means of developing sustainable fishing industries. But nobody had yet come up with a way of doing this for pearl shell. Most of the region's shallow pearl-shell beds were exhausted, and deep-sea beds could not be protected from plunderers." – Iain McCalman, *The Reef: A Passionate History: The Great Barrier Reef from Captain Cook to Climate Change*

By the end of the 19th century, the Great Barrier Reef had been thoroughly explored and the area was well on its way to colonization. In 1893, naturalist William Saville-Kent noted, "Built up by direct and indirect agencies of soft-fleshed polyps of multitudinous form and color, it flanks the Queensland coast, excepting for the presence of a few narrow intersecting channels, for a distance of over twelve hundred miles. ... The whole coast-line embraced by the Great Barrier Reef, from Sandy Cape to Torres Strait, bristles on the chart with names, conferred by that intrepid navigator, whose very titles and significance serve to distinguish them conspicuously from those of any subsequent explorer. As samples of this nomenclature, such names may be cited as those of Break-sea Spit, Thirsty Sound, Repulse Bay, Trinity Bay, Providential Channel, Possession Island, and Capes Capricorn, Upstart, Tribulation, and Flattery. All of these, and a multitude of others associated with this coast-line, carry with them a most unmistakable Cookian ring."

Saville-Kent

Saville-Kent was also one of the first authors to make reference to the ecological damage already being done to the reef as a result of human activity in the region. He explained, "The turtles and the flocks of birds referred to in Mr. Jukes' narrative have become scarce, owing to the disturbing influences of the lighthouse colony, and the extent to which the island is visited by excursionists from the mainland."

Manuel Heinrich Emha's picture of green sea turtles

At that time, however, those creatures living under the sea were still safe and well-protected. Kent observed, "Fish...of both useful and ornamental kinds, abound within and without the margin of the reef. That species of bream, Pagrus unicolor, popularly known as the Schnapper, and rightly reckoned to be one of the finest of Australian food fishes, is particularly plentiful on the banks of the north-east side of the reef, and is the special subject of attraction to visitors. The reef-pools also teem with brilliantly colored small fishes, including notably the beautiful ultramarine-blue Labroides, with yellow fins; ...also the little black fish of the same genus, decorated with a single broad, pale peacock-green, stripe...Large blue-spotted sting-rays, Myliobatis australis, bask lazily in the intervening sandy patches; and among the deeper pools the bizarre tobacco-pipe fish, Fistularia scrrata...may frequently be met. This fish in life is of a rich golden-brown hue, decorated along the sides with brilliant azure-blue spots. As may be surmised from its external features, it is a feeble swimmer, and can be easily cornered and captured. The species of Beche-de-mer observed by the author on this reef...included the valuable commercial variety known to the trade as 'Barrier Surf-Red' whose technical title, as identified by Professor F. J. Bell, is Actinopyga manritiaiia."

Picture of a Surf Redfish

Leonard Low's picture of a Balistoides viridescens

In addition to the lighthouse mentioned above, there were also three lightships stationed among the Great Barrier Reef Islands by the end of the 19th Century. While they protected the reef by warning ships away, they also harmed it by enabling people easier access to the coral. According to Kent, "Three lightships are stationed along the course between Cape Melville and Cape Grenville, to guide the passage through the intricate maze of reefs and shoals. The first, known as the Piper Islands Lightship, is a mile or so off Cape Melville. The Claremont Isles Lightship, the second, occupies a position a little less than ten miles off the mainland, about half-way between the two capes named. The most northern one, Piper Island, is twelve miles due south of Cape Grenville. Both the last-named stations have yielded specimens of interest. From the neighborhood of the Claremont Lightship, more particularly, varieties of coral which have not been collected elsewhere, have been obtained by the lightship keepers, Mr. and Mrs. Wilson. Among these is Madrcpora ornata, a new species…. The living tints are…a brilliant grass-green, with whitish terminal corallites. It has, so far, been collected only from a depth of two or three fathoms, and with the aid of native divers. From the neighborhood of the Claremont and also the Piper Lightships, some excellent quantities of sponges belonging to the honeycomb, Hippospongia, and so-called finest Turkish, Euspongia, generic, types have been obtained."

Like Jukes, Kent came to appreciate the beauty that many of the reefs offered, while also recognizing their fragility: "The most luxuriant banks of growing coral are found on the least weather-exposed, or lee, sides of the reefs …. Nevertheless, many of the apparently easily injured species, such as the delicate vase-like coralla oi Madrepora surcidosa, are found flourishing beside the most robust forms amidst the weather-side breakers at lowest tide-mark. … Luxuriant as is the growth of coral in many of the reef-scapes…this luxuriance is much exceeded on sheltered portions of the reefs that are permanently submerged. Their sloping edges, down to a depth of three or four fathoms, as seen on a calm day over the boat's side, often reveal terrace upon terrace, or literally hanging gardens, of coral growth of every variety of form and color. Specifically, these submerged corals do not differ materially from the types accessible on the surface or near the edges of the reefs at extremely low spring-tides, although in these more sheltered and permanently submerged positions they usually exhibit a more exuberant growth. In different localities, or separate portions of the same reefs, the dominating representatives of the more distinct specific types are as prevalent as on the tidally-exposed areas illustrated. Thus, one almost perpendicular bank may be completely covered with the spreading vasiform coralla of Madrepora surcidosa or pcctinata, usually of a pale-lilac or pink-brown hue, with pale-primrose or flesh-pink growing edges. Another submarine reef is as completely clothed with the brilliant rose-pink, minutely divided, clumps of Seriatopora hystrix. A third bank may include robust branching Stags'-horn varieties, resplendent with intermingling tints of electric-blue, grass-green, and violet, and comprising such specific forms as Madrepora grandis, laxa, dccipicns, and arbiiscitla. Over a very large extent of the submerged reefs, the comparatively solid, smooth-surfaced, and more or less hemispherical, coralla of the Astrasaceae and Poritidffi monopolise the growing space, to the exclusion of the branching species; almost every gradation of intermixture may obtain."

Sadly, Kent noted that hunting for the often rare birds found on the reef, or even just their eggs, was already popular sport among tourists. Without understanding the long term harm that was taking place, he wrote, "One of the strongest attractions to ordinary passengers favored with an opportunity of landing on the Cairn Cross, Howick, or other of the numerous coral-islet groups scattered along the steamer route, is the chance of making a bag of the famous Torres Strait pigeons, Myristicivora spilorrlioa, a large white variety, highly esteemed for the table, which, arriving from the north, is distributed from October until the end of March throughout the tree-bearing islets and mainland coast as far south as Keppel Bay. The nests of this pigeon are usually built in the forked-branches of the mangrove and tee trees, that form such extensive thickets along the coast-line, and each contains two white eggs. A novel spectacle to the European traveler landing on these islands may probably be afforded by his first acquaintance with the nests of the Australian jungle fowl or scrub hen, Mcgapodiiis tituinliis. These consist of huge mounds of dead leaves, grass, sticks, mold, and shells, scratched together by the adult birds in a well-shaded and sheltered situation among the Hibiscus or other bushes. The dimensions of the nest-mounds may be as much as twenty feet or more in diameter, and from ten to fifteen high, several pairs of birds commonly joining in their construction. When the mounds are completed, the birds burrow holes in the center of them and deposit their eggs, which are then left to hatch by the moist heat engendered by the decaying vegetation. As many as forty or fifty eggs, usually of a brown or brick-red color, as large as those of a turkey, are sometimes found in the largest mainland nests. The eggs, as well as the parent birds, are excellent eating. An attractively plumaged bird, very plentiful in Cairn Cross and on other of the northern Barrier islets, is the Australian bee-eater, Merops oniafns."

A picture of Torresian Imperial pigeons

Brett Donald's picture of a rainbow bee-eater

Of the latter, another explorer, A. R. Wallace, wrote, "This elegant little bird sits on twigs in open places, gazing eagerly around, and darting off at intervals to seize some insect which it sees flying near, returning afterwards to the same twig to swallow it. Its long, sharp, curved bill, the two long narrow feathers in its tail, its beautiful green plumage, varied with rich brown and black, and vivid blue on the throat, render it one of the most graceful and interesting objects a naturalist can see for the first time."

Chapter 5: That Very Partnership

"And in fact, nature offers us a model in this regard: the magical, reef-creating symbiosis between microscopic algae and a tiny polyp. We might see the algae as the heart that generates the energy its partner depends upon, and the polyp as the mind, having a purposive direction to build their joint production of coral. And also to protect it in the face of torrential forces of destruction— breakers, coral-feeders, cyclones, sediment, pollution, mining, overfishing, and water that is too hot and too acidic to bear. It is a symbiosis which, as we have seen, has survived for some 240 million years, but which will split should those harsh forces so dictate. If anything can inspire us to prevent this, it's that very partnership itself, between two of the tiniest and most fragile creatures in the sea." – Iain McCalman, *The Reef: A Passionate History: The Great Barrier Reef from Captain Cook to Climate Change*

By 1923, The Great Barrier Reef Committee had been established, "composed of representatives of the various scientific institutions throughout Australia and New Zealand, and has been formed with the object of investigating the problems—both purely scientific and economic — of one of the most unique physical features possessed by Australia." Writing for the committee, Charles Hedley outlined what was needed for the reef to survive and thrive: "Two conditions are particularly required for the existence of reef-building corals. In the first place the seawater must be pure; especially must it be free from mud; and in the second place it must be very warm. Though corals predominate in the upbuilding of a coral reef, a large share in the work is contributed by other things such as shells, foraminifera, and seaweeds."

Hedley then went on to give the modern world perhaps its best written description of the reef to date: "That portion of the beach of a coral island which is exposed by ordinary tides is paved with a litter of dead and broken fragments of coral. These are strewn in confusion and often become consolidated into rock. The living reef whose wastage builds the dry land is only uncovered by an extremely low tide. It is best viewed through a water telescope when drifting in a boat, and it presents a beautiful spectacle comparable to an -exquisite garden. In color, brown predominates, but there are brilliant touches of orange, blue, red, and green. The submarine landscape is diversified by various shapes — huge hemispheres, flat tables, many-pointed antlers, shrubs, fans, and vases. Through these coral groves swim fish of quaint form and as brightly colored as butterflies. Here and there perch sea-urchins with long spines like knitting needles or short ones like pencils. Starfish, some blue as the sky; beche-de-mer, some black as ink, are scattered about; great clams with jaws a yard across gape for the unwary. As the boats drift by, the visitor is fascinated by the wonderful shapes and colors reeled off beneath him. So vast is their population that no naturalist in the world can catalogue the furniture, or appraise the wonder, of a coral reef alive and in full bloom."

Ironically, Hedley wrote extensively about the economic uses of the reef, the kinds of activity that would soon threaten its very existence. "But the Barrier Reef has other attractions; it is the scene of several important industries which may under good management attain to larger dimensions. ... Food fishes abound in great variety and large numbers, and no doubt in the future a great trade in these will be developed. ... For many years the most lucrative business has been that of pearling. The pioneers found great pearl-shells growing on the beaches at low water; but such shell as could be reached by wading was soon exhausted. Natives of Torres Strait were then engaged to gather by diving the shell which grew in shallow water. As the more accessible fields became exhausted, shell was followed into deep and then deeper water. Finally the divers adopted the helmet diving dress and air-pump. Thus equipped the modern pearl-sheller descends from a lugger and walks about on the seafloor, picking up what shell he finds and giving directions to his crew with a signal cord held in his hand. The shells are the -property of the owner but the pearls themselves are the perquisite of the divers. From time to time the beds become exhausted by over-fishing, and the pearling fleet give them time to recover by moving to another district."

Picture of a striped surgeonfish along the reef

Picture of a starry puffer

Hedley then went on to describe how some of the reef's unique creatures were being harvested for human use. "Second in importance is the trepang or beche-de-mer industry. This creature is a sea-slug, shaped like a cucumber, about a foot long and two inches broad. It crawls about in the pools of the reef and is gathered by hand, but in deeper water diving is necessary. There are a dozen or more different kinds varied in shape and color; some of these are of no commercial value, but the higher grades are worth three or four times as much as others. The beche-de-mer are prepared for market by being boiled, gutted, and dried. At this stage the product appears like scraps of old leather. It is bagged up and sent to China, where it is esteemed as a basis of soup. … A minor industry of recent development is the gathering of trochus shell. This is a pyramidal sea-snail, of three to four inches in diameter, striped with white and crimson bands and richly nacreous within. The shells are gathered by hand upon the coral reef; the contents are removed by boiling and picking or by rotting out in the sand. They are bagged and exported to Japan, where they are cut by machinery into pearl-shell shirt-buttons."

One of the reef's life forms that nearly went extinct at the hands of humans was the humble sponge, though in writing about it, Hedley was dismissive of its usefulness: "The only commercial sponge yet found on the Barrier Reef is large and coarse, unfit for use as a bath sponge, but serviceable for such rough work as cleaning cars and machinery. It grows in large black masses on the reef-flats, and when cut open looks like a slice of bullock's liver. Perhaps a higher grade of sponge may be found or introduced and cultivated in the future; the conditions under which high-class sponges are produced in Florida seem to be repeated in Queensland."

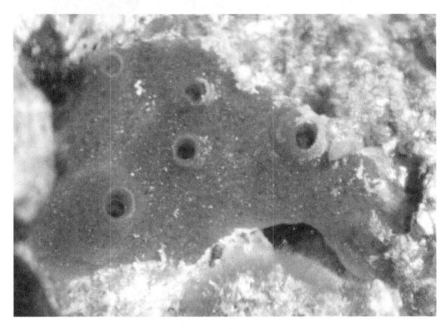

Maja Adamska's picture of a humble sponge

Perhaps the most shocking comments he made were in reference to the large sea turtle and the dugong (which is closely related to the manatee): "Both the green and the hawk's-bill or 'tortoise-shell' turtle are plentiful, though neither yet ranks as a commercial product. About November the female comes to the sandy beaches of the coral islands and digs a hole with her flippers in the dry sand above high-water mark. Here a quantity of eggs are buried and left to be hatched by the heat of the sun without further maternal care, and the young under cover of darkness escape to the sea. The egg is about the size and shape of a small billiard-ball, and though good for making cakes and puddings has the peculiarity that it can never be boiled hard. The green turtle itself supplies the favorite dish of aldermanic banquets. From the carapace of the hawk's-bill is obtained the tortoise-shell of commerce. ... The natives obtain a large supply of meat from...the dugong. This...is a marine mammal with ivory tusks, and as large as a cow; it browses on the sea-grass. The flesh has been compared to pork, and dugong oil has been recommended for lung complaints. The dugong is too rare to provide meat for the butchering trade."

A picture of a dugong

Within just a few decades of Hedley making these observations, the Great Barrier Reef was in danger of disappearing from the face of the earth, in large measure due to the fact that the very economic activities he described so well were threatening to annihilate the reef and the creatures that called it home. Fortunately, the Australian government intervened in 1975 with the passage of the Great Barrier Reef Marine Park Act. It read in part:

"(1) The main object of this Act is to provide for the long term protection and conservation of the environment, biodiversity and heritage values of the Great Barrier Reef Region.

(2) The other objects of this Act are to do the following, so far as is consistent with the main object:

(a) allow ecologically sustainable use of the Great Barrier Reef Region for purposes including the following:

(i) public enjoyment and appreciation;

(ii) public education about and understanding of the Region;

(iii) recreational, economic and cultural activities;

(iv) research in relation to the natural, social, economic and cultural systems and value of the Great Barrier Reef Region;

(b) encourage engagement in the protection and management of the Great Barrier Reef Region by interested persons and groups, including Queensland and local governments, communities, Indigenous persons, business and industry;

(c) assist in meeting Australia's international responsibilities in relation to the environment and protection of world heritage (especially Australia's responsibilities under the World Heritage Convention).

(3) In order to achieve its objects, this Act:

(a) provides for the establishment, control, care and development of the Great Barrier Reef Marine Park: and

(b) establishes the Great Barrier Reef Marine Park Authority; and

(c) provides for zoning plans and plans of management; and

(d) regulates, including by a system of permissions, use of the Great Barrier Reef Marine Park in ways consistent with ecosystem-based management and the principles of ecologically sustainable use; and

(e) facilitates partnership with traditional owners in management of marine resources; and

(f) facilitates a collaborative approach to management of the Great Barrier Reef World Heritage area with the Queensland government.

The Great Barrier Reef Marine Park would soon encompass the vast majority of the reef and therefore bring it under legal protection. Russell Reichelt, the head of the Park, recently wrote, "Our fundamental obligation is to protect the Great Barrier Reef Marine Park and the World Heritage Area. We do this by striving to ensure all human uses of the Park are ecologically sustainable and that the ecosystem's natural functions, especially resilience, are maintained. The

Reef is a phenomenal wonder and presenting its natural values to the world through industry and public uses is an important and positive part of this presentation. Sustainable tourism on the Great Barrier Reef is especially important as a way of presenting the Reef to the world. In the three decades since the Great Barrier Reef Marine Park was created there have been many changes. The number of visitors to the region has steadily escalated, as has the number of people living and working along the coastal region. We have seen waves of crown-of-thorns starfish outbreaks, highly destructive cyclones and, probably the most concerning, a rise in the frequency and extent of coral bleaching caused by increasing peak summer temperatures. ... Strong policies need to be put in place to ensure that natural, cultural and social values are adequately maintained while supporting sustainable use and long-term protection. In the case of conflicting uses, or restrictions on incremental change in use, we recommend limits to the uses of the Marine Park. Key issues for the Reef now are the effects of climate change and declining water quality, commercial and recreational fishing pressures, ports and shipping and coastal development. Our challenge is to assess, advise on, and implement policies to ensure the cumulative effects of all these issues are not leading towards a long-term decline in the environmental quality of the Great Barrier Reef."

Bibliography

Bowen, James; Bowen, Margarita (2002). *The Great Barrier Reef : History, Science, Heritage.* Cambridge : Cambridge University Press.

Hopley, David; Smithers, Scott G.; Parnell, Kevin E. (2007). The Geomorphology of the Great Barrier Reef: Development, Diversity, and Change. Cambridge University Press.

Hutchings, Pat; Kingsford, Mike; Hoegh-Guldberg, Ove (2008). *The Great Barrier Reef: Biology, Environment and Management.* CSIRO Publishing.

Mather, P.; Bennett, I., ed. (1993). *A Coral Reef Handbook: A Guide to the Geology, Flora and Fauna of the Great Barrier Reef* (3rd ed.). Chipping North: Surrey Beatty & Sons Pty Ltd.

Made in the USA
Las Vegas, NV
01 February 2025

17341098R00036